Crayfish

This book has been reviewed
for accuracy by
David Skryja
Associate Professor of Biology
University of Wisconsin Center—Waukesha.

Library of Congress Cataloging in Publication Data

Pohl, Kathleen.
  Crayfish.

  (Nature close-ups)
  Adaptation of: Zarigani / Jun Nanao, Hidetomo Oda.
  Summary: Text and photographs describe the life
cycle, behavior patterns, and habitats of crayfish.
  1. Crayfish—Juvenile literature. [1. Crayfish]
I. Nanao, Jun. Zarigani. II. Title.
QL444.M33P59 1986   595.3'841   86-26234

ISBN 0-8172-2718-0 (lib. bdg.)
ISBN 0-8172-2736-9 (softcover)

This edition first published in 1987 by Raintree Publishers Inc.

Text copyright © 1987 by Raintree Publishers Inc., translated by
Jun Amano from *Crawfish* copyright © 1977 by Jun Nanao and
Hidetomo Oda.

Photographs copyright © 1977 by Hidekazu Kubo.

World English translation rights for *Color Photo Books on Nature*
arranged with Kaisei-Sha through Japan Foreign-Rights Center.

 2 3 4 5 6 7 8 9 0   90 89 88

Crayfish

Adapted by
Kathleen Pohl

Raintree Publishers
Milwaukee

▲ **Crayfish coming out of their burrows.**

Crayfish burrow into the ground to escape the winter cold. Their body functions slow down. In spring, they crawl out of their holes and become active again.

The crayfish is not really a fish at all. It is actually a close relative of the lobster. Both crayfish and lobsters belong to a large class of animals which scientists call crustaceans. Crustaceans are animals whose bodies are protected by hard shells. Crabs and shrimps are also crustaceans.

Crayfish live in lakes and rivers throughout the world except in Africa and the Antarctic. In the United States, there are more than 230 kinds, or species, of crayfish. They vary in color and size. They may be white, pink, orange, brown, or dark blue. The species discussed in this book is red. Most crayfish don't grow larger than six inches long. But a few Australian species measure more than a foot and weigh as much as eight pounds.

Crayfish are cold-blooded animals. That means that their body temperature changes as the temperature of their surroundings changes. In winter, crayfish burrow into the mud at the bottom of ponds and rivers to escape the cold weather. When spring comes, they crawl out of the mud and become active again.

▼ A crayfish looking for water.

This crayfish is crawling out of its hole in a muddy field to look for a pond or river. Although crayfish usually live in water, they can stay on land so long as their gills stay wet, so they can breathe.

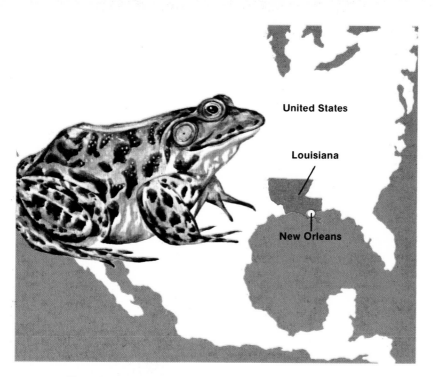

◄ Crayfish are plentiful in the United States.

In 1930, crayfish were exported to Japan from New Orleans as food for bullfrogs. The shellfish are now found in most warm parts of Japan. In the United States, crayfish abound in the bayous of Louisiana.

► A crayfish waving its pincers to scare off an enemy.

A crayfish has five pairs of legs. The first pair are large, claw-like pincers. They are used for catching food and for self-defense. The other four pairs are walking legs.

Crayfish are eaten by people in many countries, including the United States. In the southern United States, crayfish are called "crawfish" or "crawdads." In Louisiana, crayfishing is a major food industry. Millions of pounds of crayfish are harvested there each year. The crustaceans thrive in Louisiana's wet marshes and bayous.

Crayfishing has also become an important tourist industry in Louisiana. Local festivals each year include contests to see who can shell the most crayfish, and who can eat the most shellfish in a certain amount of time. There are even contests to see whose crayfish can crawl the fastest.

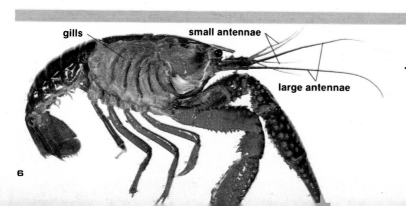

◄ The gills of a crayfish.

Crayfish breathe with gills, just as fish do. The gills take in oxygen from the water. Special sensory hairs on its antennae help the crayfish find food.

◄ A crayfish protecting its hiding place.

Sometimes two crayfish fight over the same hiding place. But they don't usually fight as seriously with one another as they would with real enemies. Usually the crayfish with the larger claws wins.

Crayfish do not like warm weather and sunshine. So they hide behind stones or burrow into the mud during the day. They come out at night to hunt for food. Special sense hairs on their antennae help them to smell food nearby. Crayfish eat tadpoles, snails, small fish, and insects. They use their large, clawlike pincers to grasp their prey. Crayfish also eat plants and waterweeds.

If another animal comes too close to a crayfish's hiding place, the crayfish will raise its claws to defend itself. The crayfish is also protected by its hard, shell-like covering, called the exoskeleton. This acts like a suit of armor to protect the soft, inner body tissues. A special shield, called the carapace, protects the crayfish's head and midsection, or thorax.

◄ Ink was added to water to show how a crayfish takes in and expels water as it breathes.

The crayfish takes in water through its gills, beneath the carapace. It breathes oxygen from the water, then discharges the water through openings near the front of the carapace.

▼ **Two crayfish threatening one another.** Crayfish lift their pincers in a menacing way when they feel threatened. By doing so, they try to frighten their enemies.

The crayfish uses one of its walking legs to shower itself with sand. Some sand goes into the crayfish's "ear" openings. Scientists believe this helps it to keep its balance.

An "ear" opening at the base of the smaller antennae.

All crayfish have the same basic body structure. The head and thorax together are called the cephalothorax. They are protected by the carapace. The back section of the body is the abdomen. It is commonly called the tail.

Crayfish have five pairs of legs attached to the thorax. In addition to the first pair, the clawlike pincers, there are four smaller pairs that are used for walking. Crayfish can walk forward, backward, and sideways.

They also use their walking legs to help tear food apart and to clean themselves.

The crayfish has two sets of feelers, or antennae. At the base of the smaller antennae, there are tiny "ear" openings. These are lined with sensory hairs. Crayfish cannot hear, but they can detect vibrations in the water. The antennae and pincers are also covered with tiny hairs, which are sensitive to touch, smell, and taste.

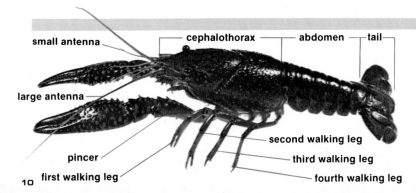

small antenna
cephalothorax — abdomen — tail
large antenna
second walking leg
pincer
third walking leg
first walking leg
fourth walking leg

◄ **The crayfish's legs.**

Notice that the first two pairs of walking legs also form tiny claws. These are useful in tearing apart food.

▼ A crayfish using its small claws to clean its tail.

Crayfish clean their bodies so they don't become covered with mold. They can feel places to clean even if they cannot see them because their bodies are covered with tiny sensory hairs.

Crayfish eat plants as well as small animals. They also feed on decomposing plant and animal matter.

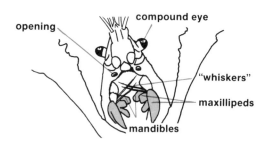

▶ Close-up of a crayfish's head.

The crayfish's eyes stand out slightly from the head on eyestalks. The compound eyes are made up of many tiny lenses. In the photo, some of the crayfish's outer mouthparts are visible.

The crayfish seizes its food with its pincers, or with the first or second pair of walking legs, which also serve as small claws. These grasp the food and begin to tear it apart. Then the food is passed on to a complex set of appendanges, or outer mouthparts. These include: a pair of true jaws, called mandibles; two pairs of maxillae; and three pairs of maxillipeds. The maxillae and maxillipeds handle the food and help to tear it into pieces. It is only the mandibles that actually chew the food. The crayfish's mouthparts move from side to side, rather than up and down, as people's jaws do.

The crayfish has a kind of mill in its stomach which further grinds up the food. It is then digested in the stomach. Waste matter is passed to the intestines.

► **A young crayfish eating a dead fish.**

Crayfish are scavengers, animals which eat dead plant and animal material.

◄ **This snowy heron has caught a crayfish.**

Water birds with long beaks, like this heron, can reach into a crayfish's hiding place to capture it.

Adult crayfish are not good swimmers. But when one is threatened by an enemy, it can flip its tail quickly and push itself backwards through the water. In this way, it can move quickly away from hunters, or predators.

Crayfish have many enemies. Opossums, raccoons, bears, turtles, fish, some birds, and even people hunt, or prey upon, crayfish.

Like crabs and lobsters, crayfish have the remarkable ability to grow new body parts. If a crayfish loses a claw, leg, or even an eye in a fight with another animal, it can replace it by growing a new one. This process is called regeneration.

◄ **A regenerated claw (left) and legs (right).**

If a crayfish has been caught by another animal and can't get free, it may detach a leg or claw in order to escape. Later, a new limb will replace the lost one.

▼ **A crayfish "swimming" through the water.** This crayfish has snapped its tail forcefully, moving it backwards through the water.

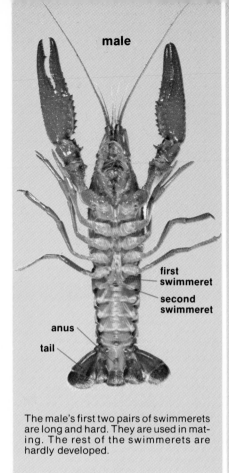

male

first
swimmeret

second
swimmeret

anus

tail

The male's first two pairs of swimmerets are long and hard. They are used in mating. The rest of the swimmerets are hardly developed.

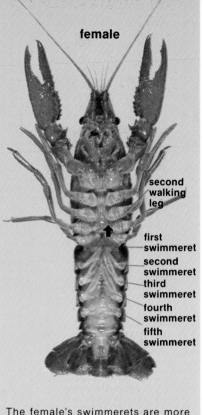

female

second
walking
leg

first
swimmeret

second
swimmeret

third
swimmeret

fourth
swimmeret

fifth
swimmeret

The female's swimmerets are more developed. At the base of the second pair of walking legs are openings through which she lays her eggs.

▶ **A male crayfish (left) chasing a female.**

At first, the female may seem to flee from the male. But soon she will respond to his advances.

▶ **Crayfish mating.**

The male has turned the female on her back so the two can mate. For this species, the mating process takes about twenty minutes.

It is not easy to tell a male crayfish from a female. But if you flip them over, you can see a difference. On the underside of the female's abdomen are five pairs of swimmerets, small leglike limbs. These are an aid in swimming, but they are mostly used for carrying her eggs. The male's swimmerets are much less developed, except for the first two pairs, which are used in mating.

It is also possible to tell a male from a female by watching their behavior. In springtime, during the mating season, male crayfish are very aggressive. They constantly chase after other crayfish. Other males will fight back, or move quickly away. But a female will not resist the male's advances. Once a male finds a female, he will flip her over on her back, and the two will mate.

◀ **A river where crayfish live.**

▶ **A cluster of crayfish eggs.**

These eggs are attached with a kind of glue to the tiny hairs on the female crayfish's swimmerets. If danger threatens, she curves her abdomen around the eggs to protect them.

During the mating process, the male crayfish gives deposits of sperm to the female. Later, the sperm will join with the eggs she carries in her abdomen. From these fertilized eggs, baby crayfish will develop.

The number of eggs laid depends on the species of crayfish. The female of this species lays from 100 to 300 eggs. She attaches them with a sticky, gluelike substance to the hairs on her swimmerets. She carries them there until they hatch. The crayfish shakes her body constantly and paddles her swimmerets so that the eggs are washed clean in the water. In this way, too, they get a supply of fresh oxygen. Inside the eggs, the bodies of the baby crayfish are developing.

◀ **A mother crayfish with her eggs.**

From the time she lays her eggs until they hatch, the crayfish spends much of her time in her burrow, hiding from enemies, to protect the eggs.

▲ A mother crayfish with her newly hatched young.

When crayfish babies are first born, they cannot live by themselves. They stay with their mother for several weeks before going off on their own.

In about two weeks, the eggs of this species hatch. As they break out of their eggshells, the tiny baby crayfish look much like their parents. They have huge eyes and large hump-shaped carapaces. But their bodies do not yet have the bright colors of the adult crayfish.

The baby crayfish are helpless at first. For a short time, they remain attached by a stringy substance to their mother's swimmerets. She beats her swimmerets back and forth to send fresh water over her young so they will have oxygen to breathe.

In the first two weeks, the baby crayfish shed their skin, or molt, several times. As they do, the string that binds them to their mother breaks. But the young crayfish still cling to her with special claws that have hooked tips.

A newly hatched crayfish still attached by a string to its mother (left) and a young crayfish after it has molted twice (right).

● How baby crayfish grow.

Within two weeks after they hatch, baby crayfish double in size. With each molting, they look more and more like adult crayfish.

▼ Baby crayfish hiding under their mother's abdomen.

If the mother senses danger, she will curve her abdomen around her babies to protect them.

▲ **A young crayfish feeding on water plants.**
This tiny crayfish's skin is still transparent. If you catch a crayfish this size, you can watch its stomach move as the crayfish eats.

▲ **A growing crayfish feeding on water worms.**
As the crayfish gets older, it begins to feed on worms it finds near the bottom of the pond.

By the time the crayfish are two weeks old, they have doubled in size. They are ready to leave their mother and go off on their own.

When the young crayfish set out on their own, they swim to shallow water. They live for a while on water plants growing at the edge of ponds or along the banks of rivers. At first, they feed only on the tender young leaves of water plants. Their pincers must grow larger before they are useful in catching prey.

As the crayfish grow older, they leave the water plants and begin to crawl about on the bottom of ponds or rivers. They search for earthworms and other kinds of water worms—which they can digest easily.

Crayfish, in turn, are preyed upon by fish and other water animals. In this way, the underwater life cycle continues. Every plant and animal contributes in some way to the food chain.

◄ **Crayfish that have just left their mother.**
These baby crayfish used their swimmerets to swim to shallow water, where they will live on water plants for a time.

● **A crayfish molting (photos 1-5).** (1, 2) The carapace splits down the back and the head appears. (3, 4) The crayfish pulls out its legs and long antennae. (5) The abdomen and tail are worked free from the old exoskeleton. The molting is complete.

As the crayfish eats and grows larger, its exoskeleton does not grow with it. So the crayfish, like other crustaceans, must shed, or molt, its shell-like covering. Young crayfish molt more often than older ones.

Before the old shell can be shed, a new one must form beneath the old one. This is the time when new limbs are regenerated, if a crayfish has lost a leg or pincer in battle.

When the crayfish molts, its carapace splits down the back. First its head, with its antennae and complex mouthparts, pokes out. Then the thorax, with the pincers and walking legs, emerges. Finally, the abdomen and tail appear.

◀ **A crayfish ready to molt.**

When a crayfish is ready to molt, it stops eating. The space between its carapace and abdomen expands. Soon the crayfish will begin to molt.

◄ **A crayfish that has failed to molt.**

The crayfish absorbs calcium into its body from the old carapace in order to soften it. But it is still difficult for the animal to work its way out of the old shell. Crayfish often fail to molt successfully.

Molting is a dangerous time for crayfish. If a leg or some other part of its body gets caught during the molting process, the crayfish will not be able to complete its molt and will soon die. Even if it molts successfully, the crayfish's new shell is soft at first, so predators can easily attack it.

Just before a crayfish molts, the calcium which makes its shell hard is absorbed into the crayfish's body.

This softens the old shell, making it easier to shed. The calcium collects in the crayfish's stomach in a hard ball, called a "crab's eye." After the crayfish has molted, the calcium in the crab's eye is dissolved by digestive juices. It is redeposited in the soft new shell and helps to harden it. But until the new shell hardens, the crayfish is easy prey—even to other crayfish.

◄ **Crab's eyes.**

Centuries ago, when people didn't understand the real purpose of crab's eyes, they were thought to be good luck charms. People believed that crab's eyes could cure illnesses.

▼ A crayfish attacking another crayfish that has just molted.

Crayfish usually stay in their burrows just before and after molting, when their carapaces are soft. If they don't, they may be attacked by other crayfish.

▲ **A crayfish digging a burrow.**

Crayfish use their claws, or pincers, to dig in the mud. With the coming of winter, crayfish dig into the earth to avoid the cold temperatures.

▲ **A burrow that has just been dug.**

This species of crayfish digs holes that are sometimes a yard long. Notice the chimney at the entrance to the burrow.

In wet areas, crayfish sometimes leave ponds and rivers to live on land. They burrow deep into the earth until they reach the water table. That way, they can keep their gills wet so that they are able to breathe. As they burrow, they pack mud at the entrance to their burrows. These mounds of mud are sometimes called chimneys. They may be anywhere from six to eighteen inches high, depending on how deep the burrow is. Burrows are usually a few inches to a yard long. But in the southern United States, one species of crayfish digs much deeper holes—up to twenty feet! Crayfish holes can cause great damage. If enough of them are dug in one place, whole riverbanks or irrigation ditches may collapse.

▶ **Crayfish burrows.**

These crayfish burrows have been filled in with mud and water. Inside them are crayfish.

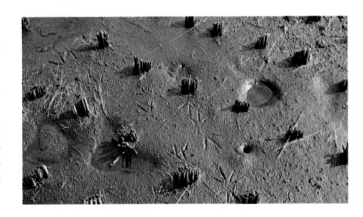

▼ **An adult crayfish walking on the bottom of a pond.**

This species of crayfish is full grown after it has lived through its second winter. Other species may take five years to mature. Although many crayfish never reach maturity, others live to be fifteen or twenty years old.

Let's Find Out

How to Catch and Raise Crayfish

▼ (**1**) A crayfish in a rice paddy in early summer. (**2**) Children catching crayfish. (**3**) A crayfish's pincers. (**4**) If you pick up a crayfish, hold it by its carapace so it will not pinch you. (**5**) A crayfish crawling among water plants.

How to Catch Crayfish

Use a net to catch crayfish. Or you can fish for them with bait tied to the end of a string. Use a piece of meat for bait. Sometimes two or three crayfish will bite at the bait at one time. In late fall, look for crayfish holes in riverbanks. Reach into the burrow to catch a crayfish. Be sure to wear gloves so you don't get pinched.

How to Raise Crayfish

A Container and a Hiding Place

Place sand and stones in an aquarium. Add water plants and pondscum for crayfish to eat. Use an air pump to add oxygen to the water. Cover the aquarium so your crayfish will not escape.

Crayfish need things to hide under.

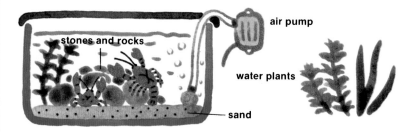

Food for the Crayfish

Feed your crayfish dried fish. They will also eat sweet potatoes, pumpkins, and earthworms. Make sure to take out the extra food so the water will not become dirty.

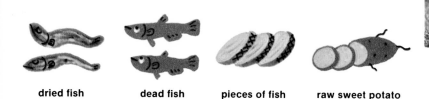

dried fish   dead fish   pieces of fish   raw sweet potato

A crayfish eating a dead fish.

How to Take Care of Young Crayfish

When the baby crayfish leave their mother, put them in a separate aquarium. Give them water plants and cooked spinach to eat. When the crayfish get older, feed them fish food.

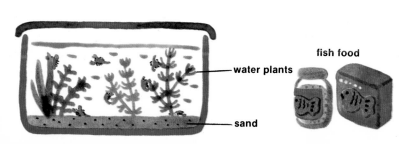

Young crayfish that have just left their mothers feed on tender water plants.

GLOSSARY

carapace—the protective shield that covers the crayfish's cephalothorax. (pp. 8, 24)

cephalothorax—the crayfish's head and midsection, or thorax, combined. (p. 10)

cold-blooded animal—an animal whose body temperature changes with the temperature of the air. (p. 4)

crustaceans—a large class of animals whose bodies are covered with hard, shell-like coverings called exoskeletons. (pp. 4, 6)

molting—the process by which crayfish and crabs shed their old exoskeletons. (pp. 20, 24)

predators—animals that hunt and kill other animals for food. (p. 14)

regeneration—the process by which a crayfish or crab grows a new claw or leg to replace one it has lost. (p. 14)

scavengers—animals that feed on dead or decaying plant and animal matter. (p. 13)

species—a group of animals which scientists have identified as having common traits. (pp. 4, 28, 29)

swimmerets—small appendages on the underside of the crayfish's abdomen, most prominent in the female. They are used in carrying her eggs. (pp. 16, 19, 20)